Designed to Infect:

Helicobacter pylori

Jesse Domingo

DEDICATION

I would like to dedicate this book to the Salma Lab at The Fred Hutchinson Cancer Research Center. They have continually supported me and have allowed me to share these great images with the public. Thank you so much!

ACKNOWLEDGMENTS

This book would not have been possible without the support and encouragement of my science teacher and mentor, Mrs. Dunn. I would also like to thank my family for their unwavering support.

The two images on the next two pages are false colored *Helicobater pylori*, obtained using a Scanning Electron Microscope.

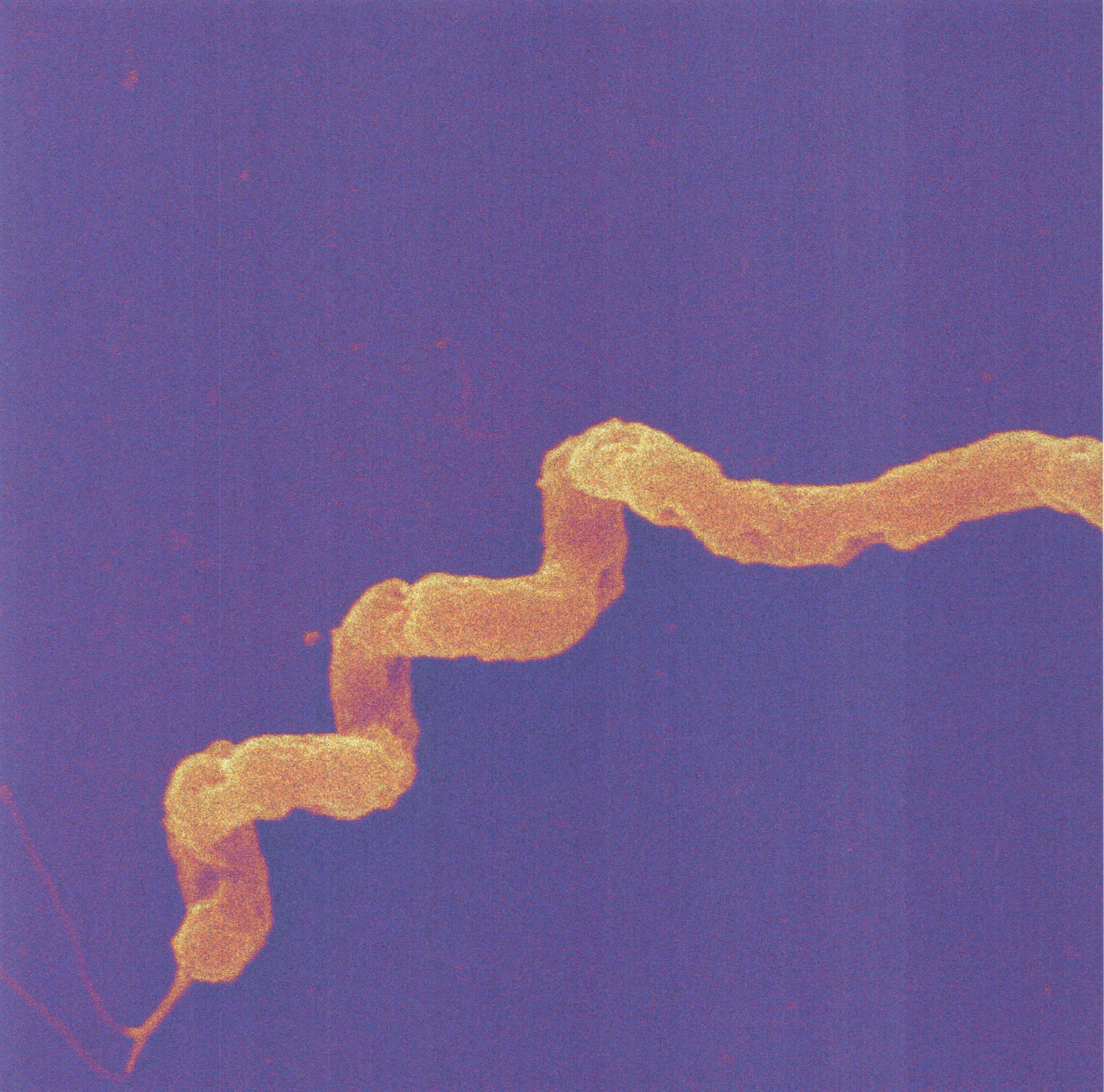

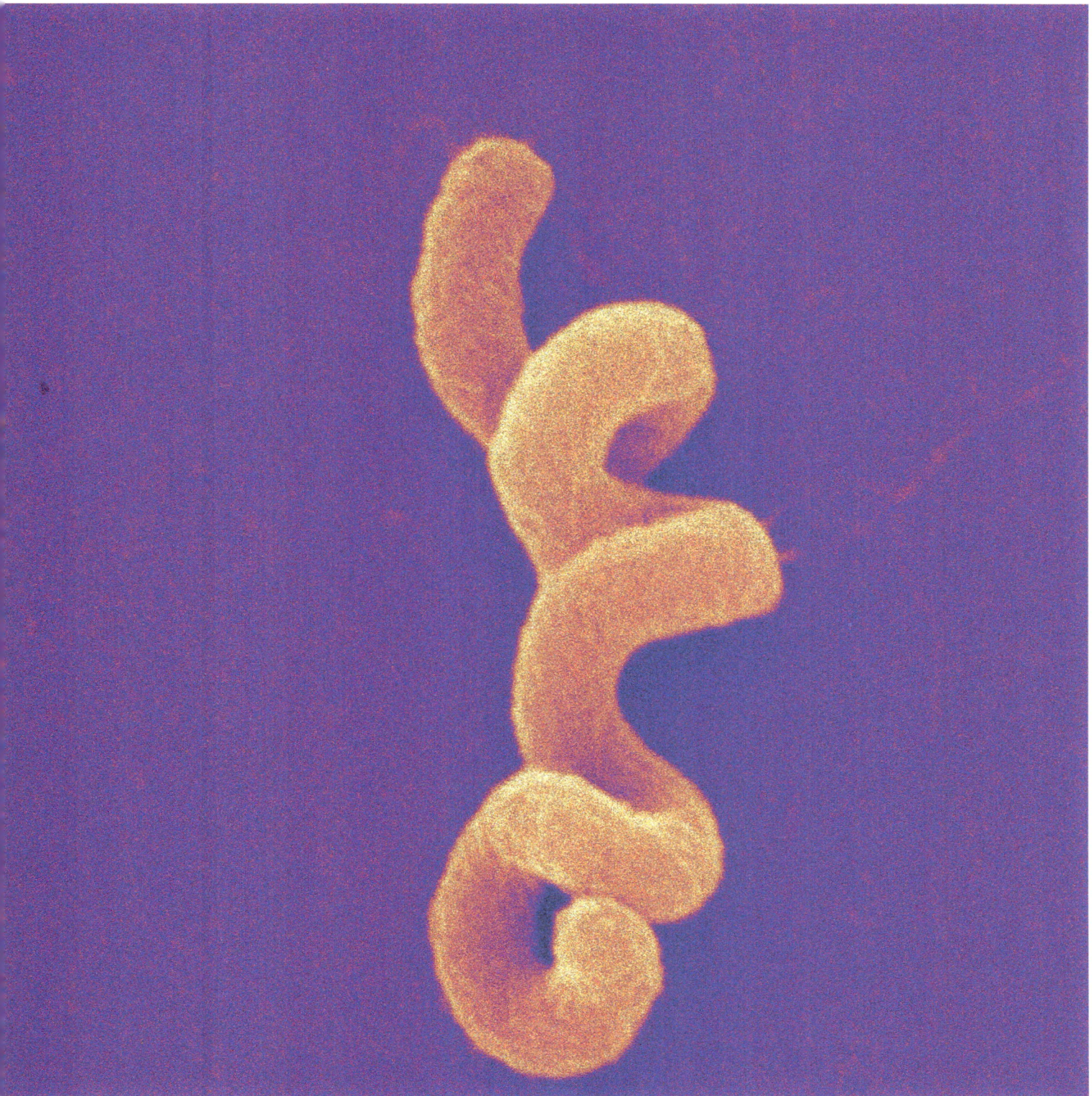

Helicobacter pylori is a bacterium that infects humans, and is normally found in the stomach. *H. pylori (Helicobacter pylori)* causes stomach ulcers and potentially stomach cancer. *H. pylori* infects more than half of the world's population. Though it is more prevalent in developing countries. The United States does not currently screen for *H. pylori* infection.

The image on the right page is of false colored *Helicobater pylori*, obtained using a Transmission Electron Microscope.

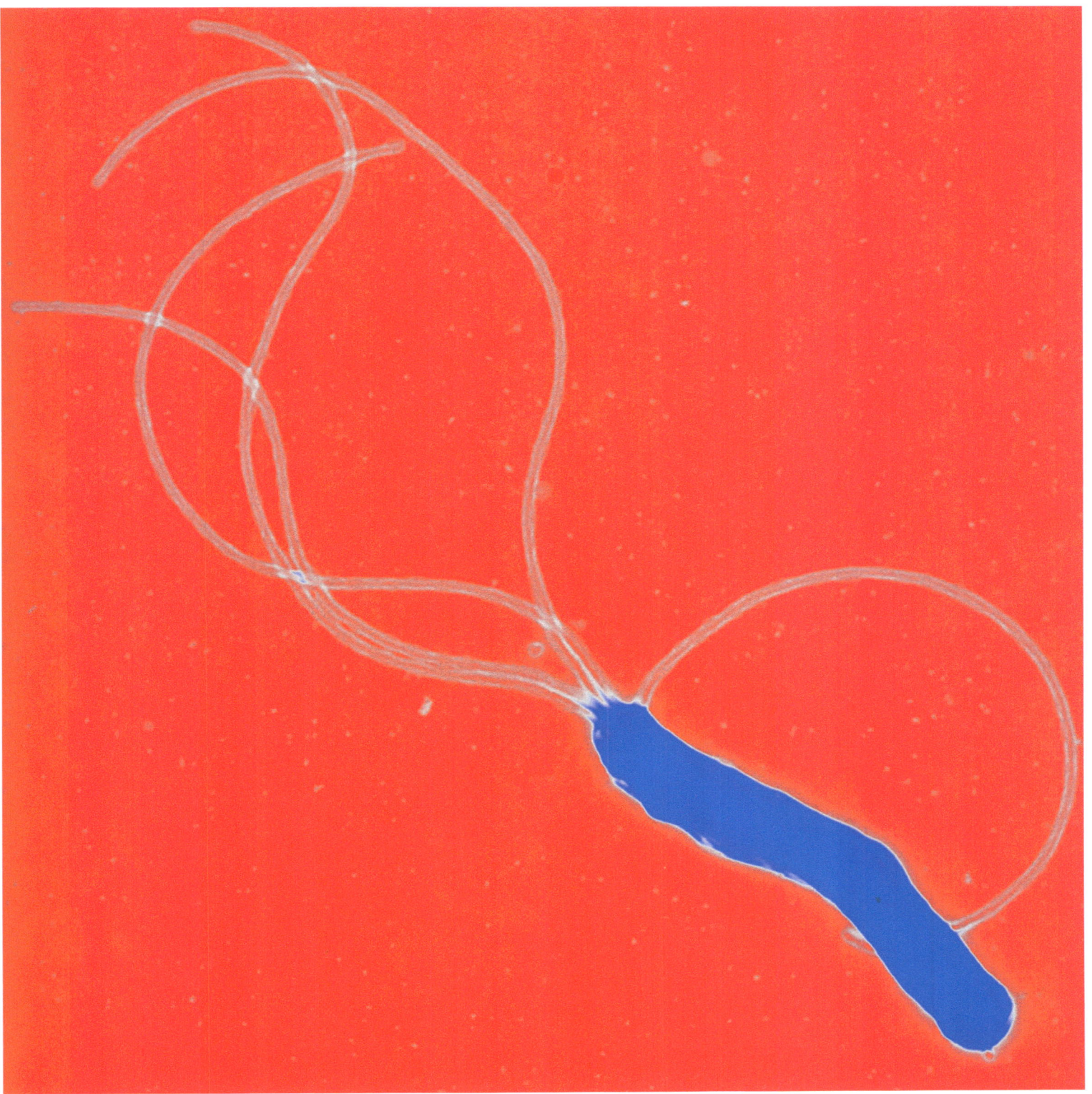

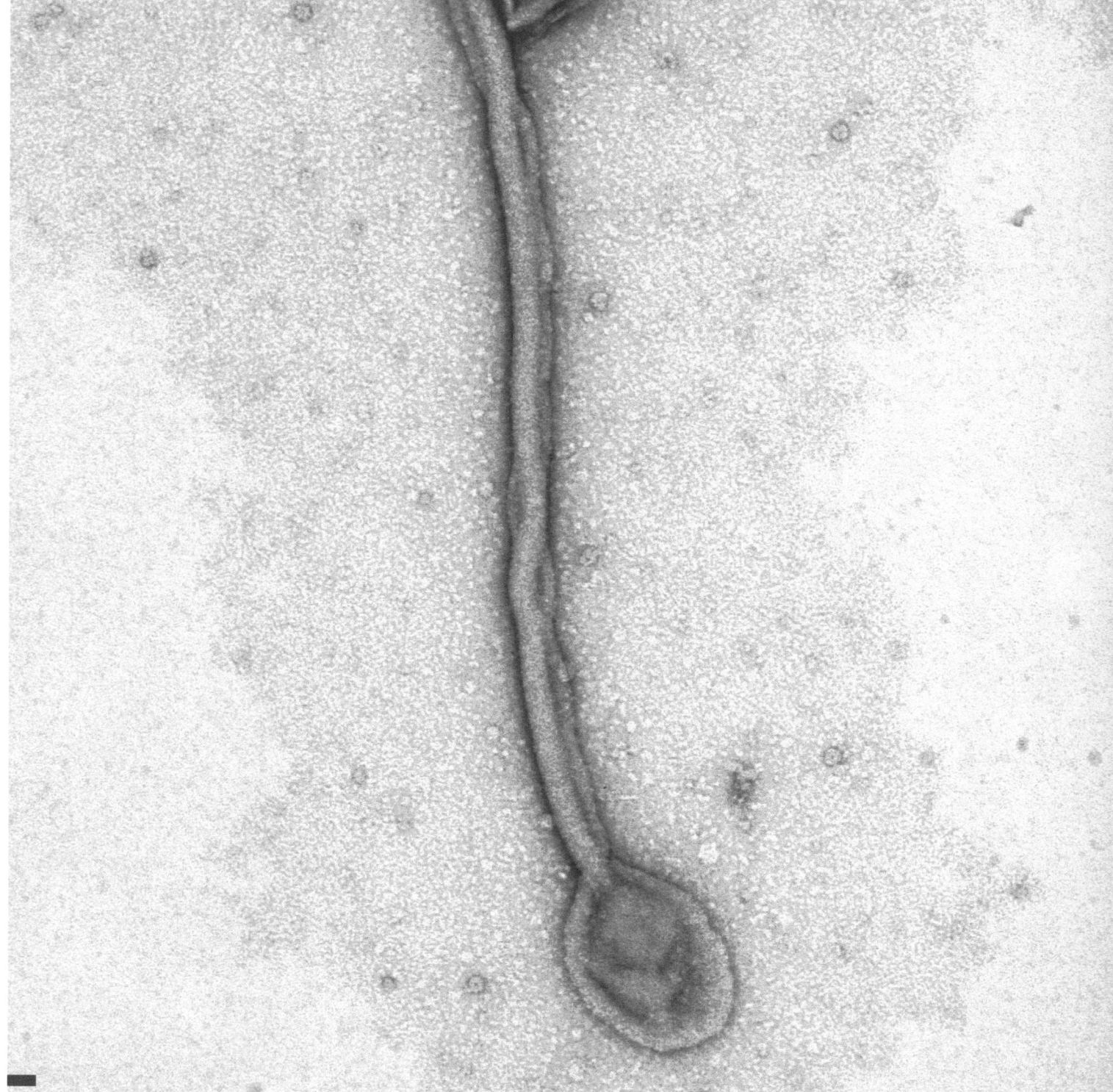

H. pylori is able
to move and find itself a
habitable zone. With its simple yet
intricate flagella it can move through the mucus
layer of the stomach with ease, to find
protection.

The image on the left page is of false colored *Helicobater pylori*
falgella obtained using a Transmission Electron Microscope.

Not only do the Flagella help with movement, but also the overall shape of the cells body assists the bacterium in finding its habitable zone. *H. pylori* has a spiral shape. It is thought that the spiral shape make it easier to swim and survive in the stomach.

The image on the right page is of mulitple false colored *Helicobater pylori*, obtained using a Scanning ectron Microscope.

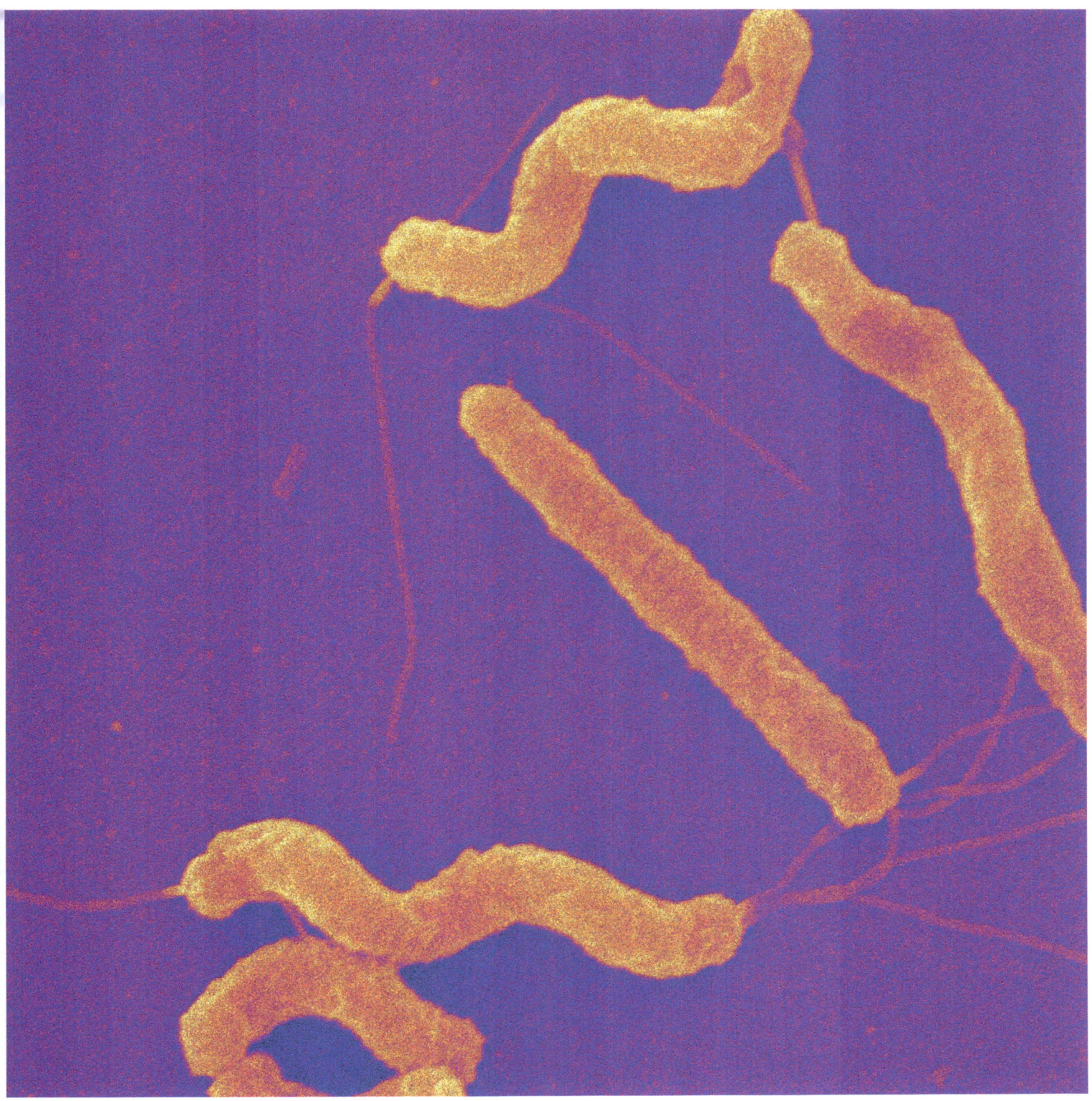

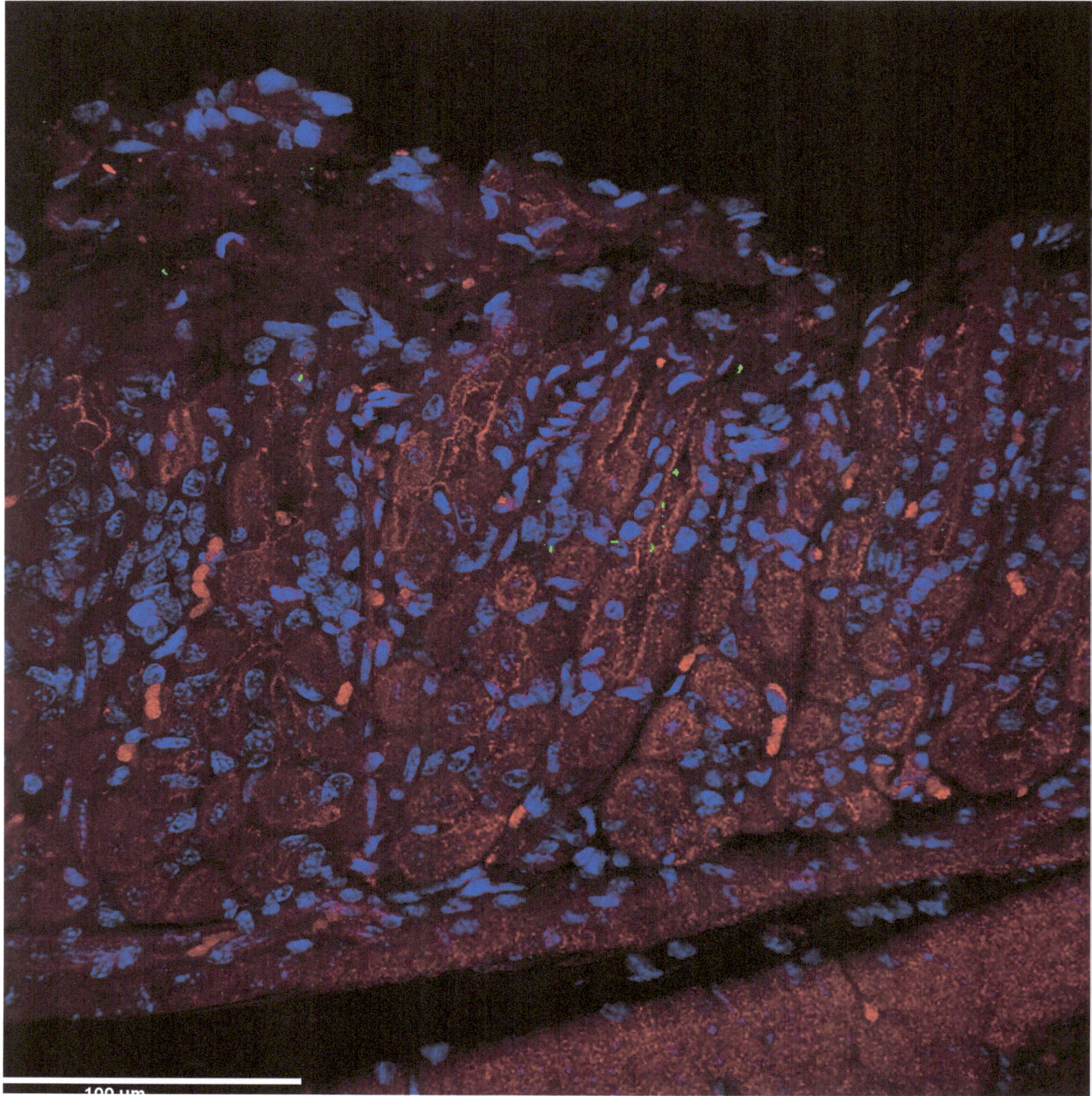

A safe spot to colonize is deep into the stomach's epithelium (lining). It is there that the bacteria can be safe from stomach acid. In the epithelium *H. pylori* will reproduce and keep spreading.

The image on the left page is of a mouse stomach infected with *Helicobacter pylori* (green is the bacteria, blue is cell DNA, and red is the cytoskelton) obtained using a confocal microscope.

Just like all biological organisms some things can go wrong in the gene code. When errors occur, mutants are formed. To the left you can see a long mutant of *H. pylori.*

The image on the right page is of mutant *Helicobater pylori.* It is colored using fluorescence staining.

The images on the next two pages are of a single mutant *Helicobater pylori* growing over a period of 18 hours. (look from right to left over the two pages.)

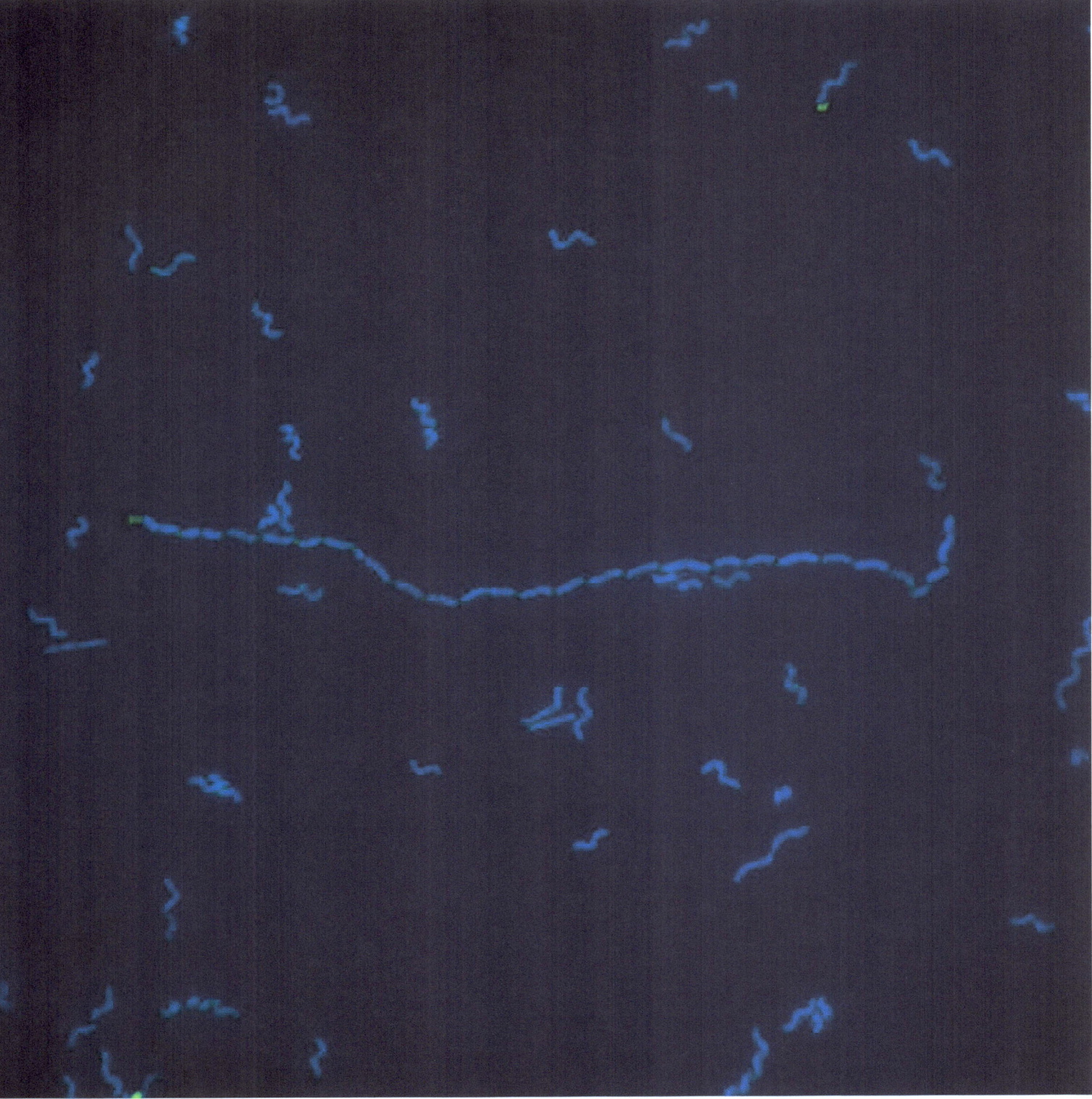

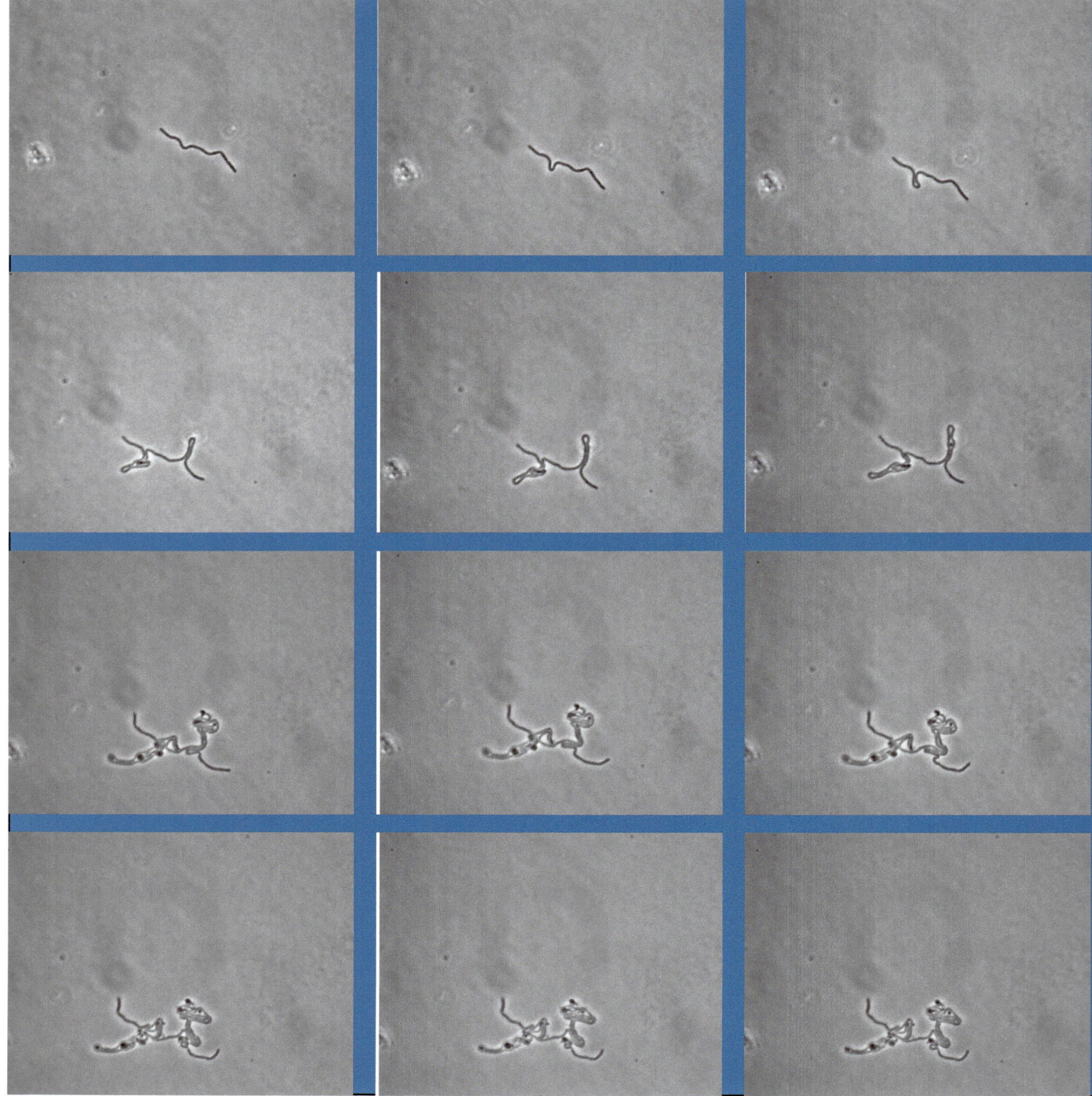

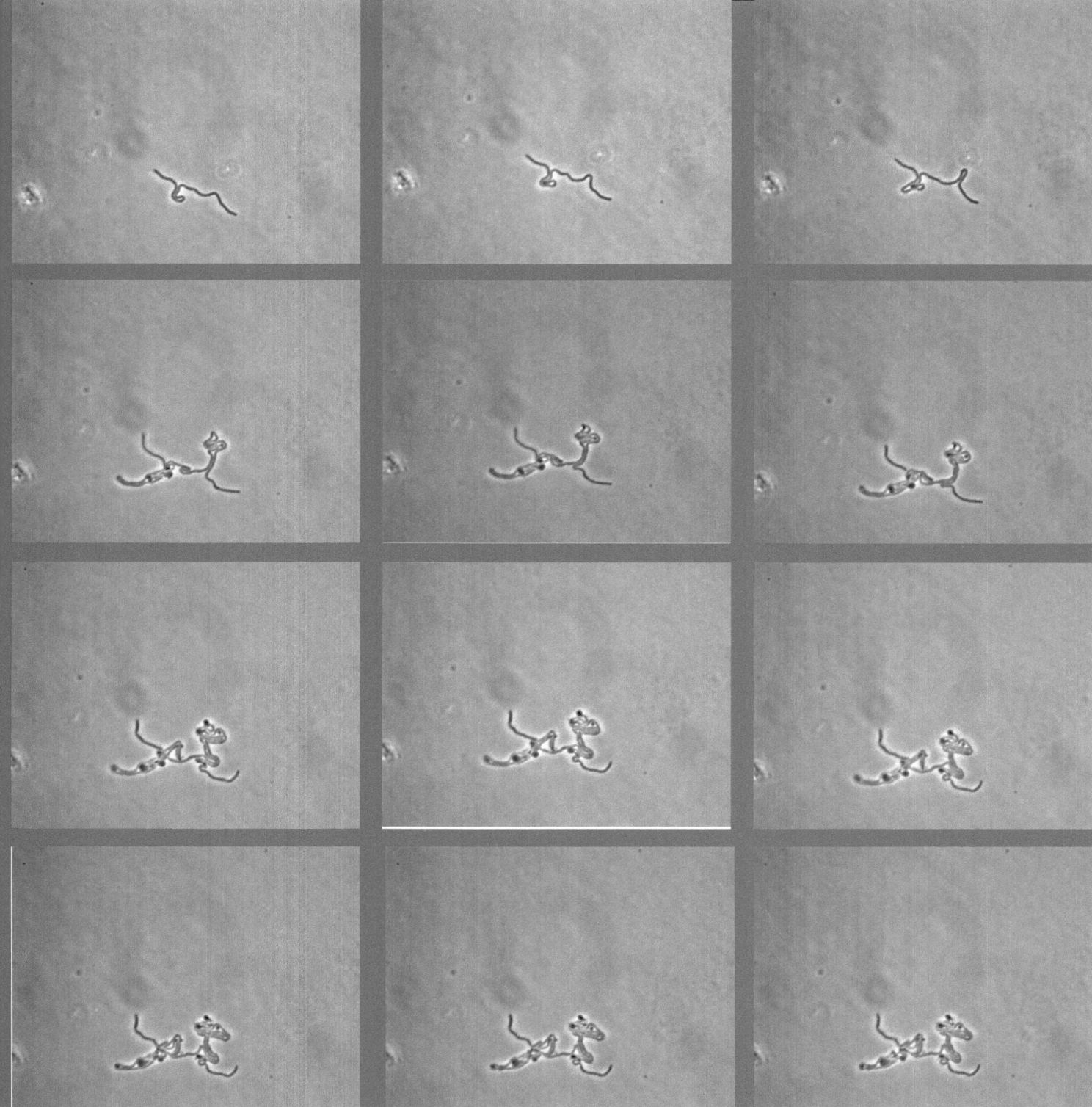

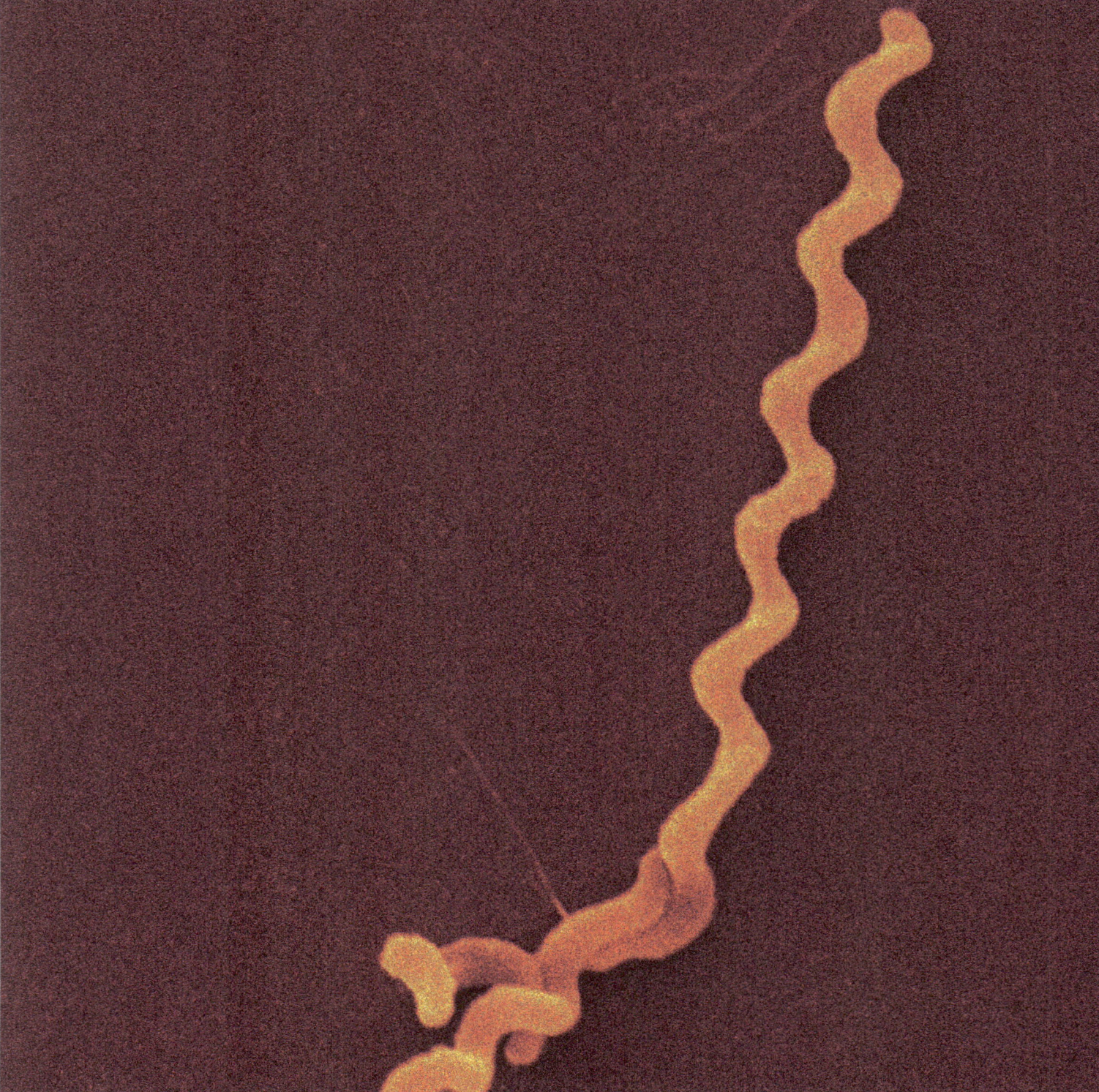

We can see in detail that some
H. pylori can have many spirals. This abundance
may help or hinder the bacterium's ability to infect.

The image on the left page is spiral mutant of *Helicobacter pylori*,
obtained using a Scanning Electron Microscope. (false colored)

On the other hand,
H. pylori also has a straight form.
Scientists are currently trying to determine
which genes activate its spiral shape. They are also asking
the question: Are bacteria mutating to accommodate for
different regions of the stomach? If so does the mutant
have an easier time colonizing because of its
shape change?

The image on the right page is straight mutant of *Helicobacter pylori*,
obtained using a Scanning Electron Microscope. (false colored)

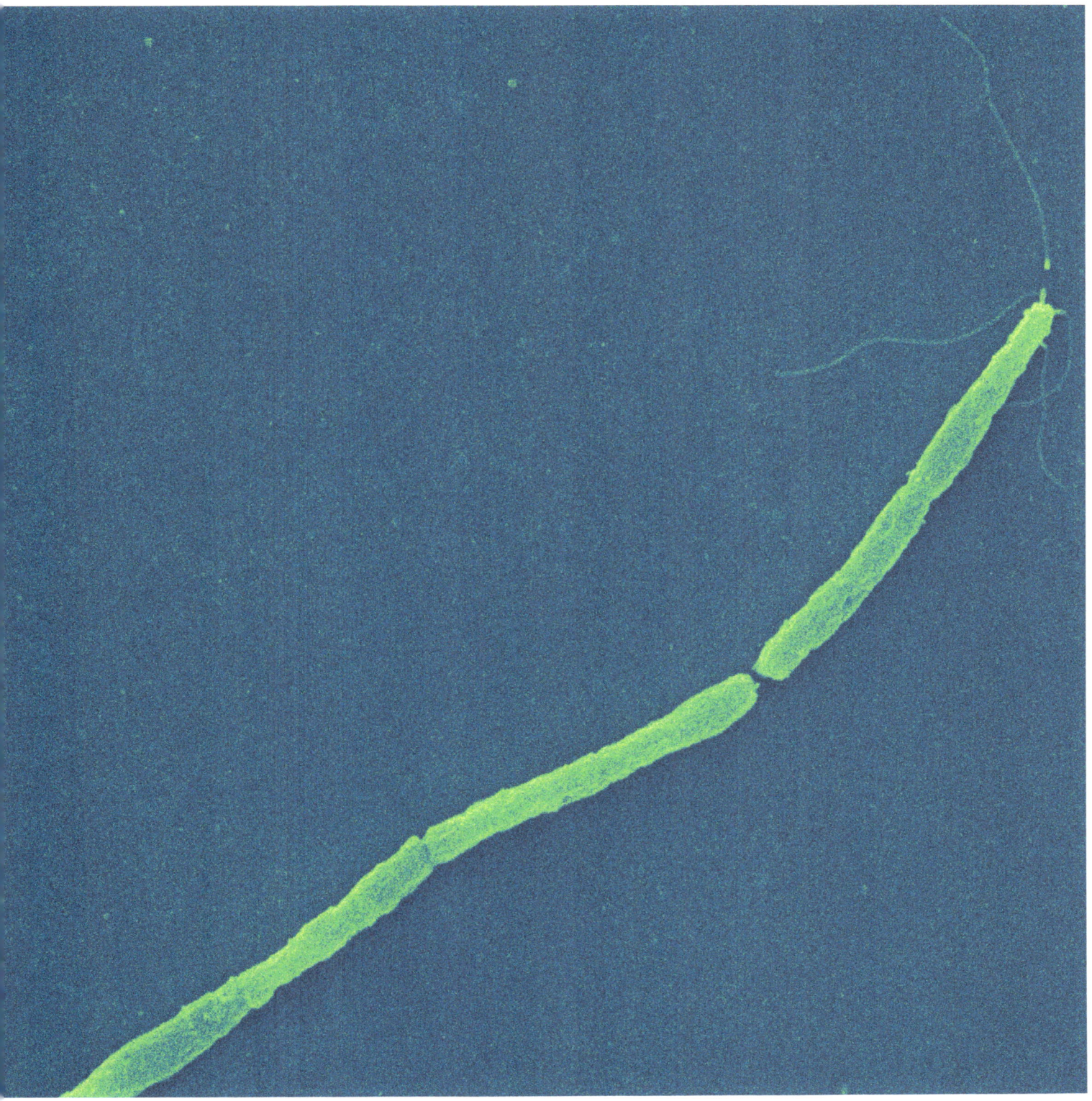

The image on the left are of
Helicobater pylori in a solution, obtained using
a confocal microscope.

Some *H. pylori* use a
needle-like appendage to inject a
toxin into the junctions where cells of he
stomach lining meet. This seems to increase the risk
of stomach cancer, which currently kills over 10,000 people per
year in the United States, and many more globally.
Whatever the case, this well-designed, intricate
killer deserves more research.

The two images on the next two pages are false colored *Helicobater
pylori*, obtained using a Transmission Electron Microscope.

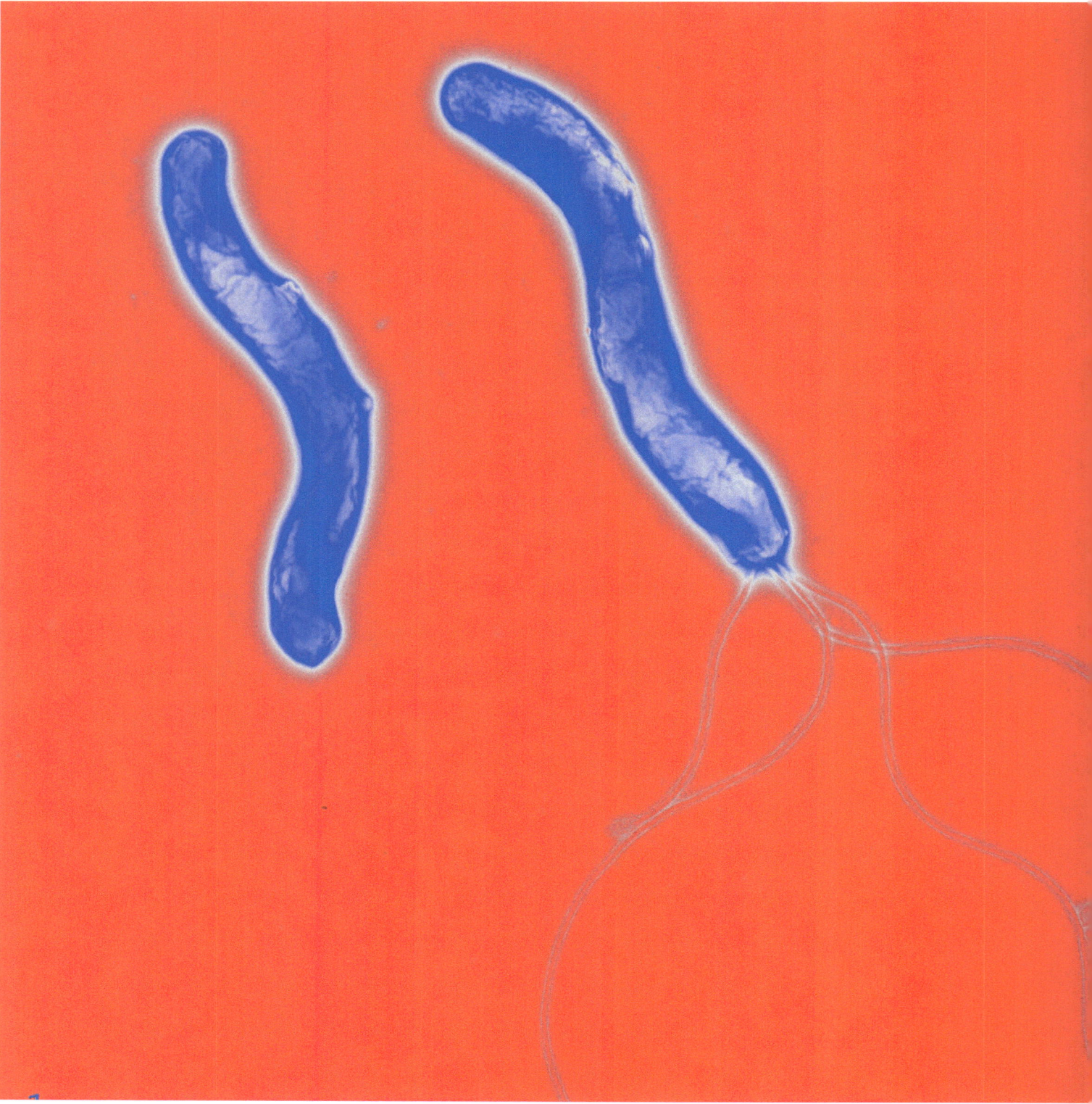

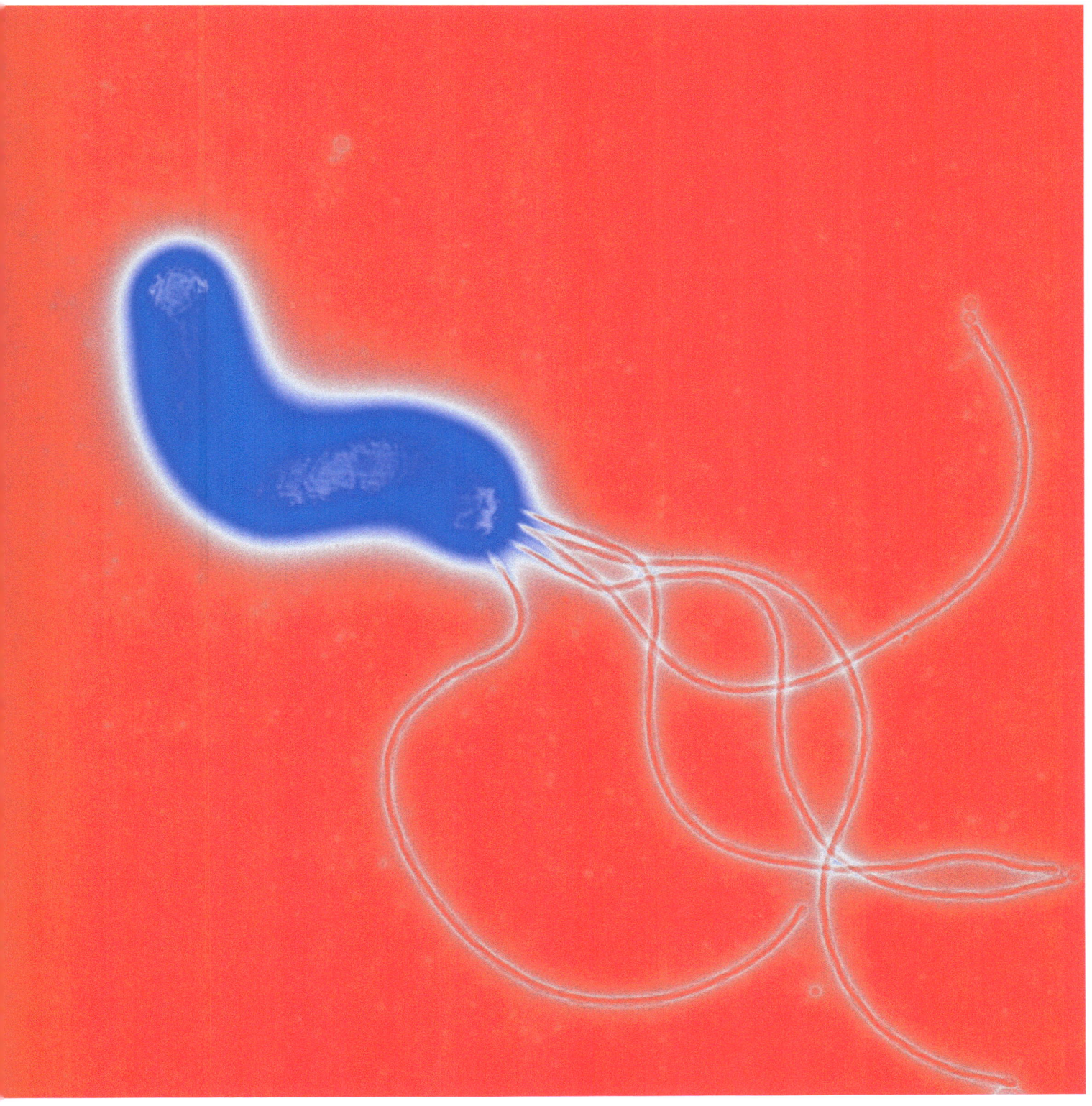